回 味 下 飯 菜

Those were the Tastes

方任利莎 著
Lisa Yam

目錄 Contents

回味下飯菜
～與您分享最受歡迎的經典菜式

這是我和圓方出版社第五次的合作。

事緣一次和編輯小姐茶敘，偶而向她提起想把以前在電視中示範過的食譜丟棄，因為實在太多了。編輯小姐說不能丟，並願為我整理。恰在此時我的烹飪節目重播，想不到多年前的節目仍能受到大批觀眾捧場喜愛，編輯小姐建議不如我們把當年受歡迎的食譜再加新的構思出本烹飪書吧。

雖說是以前的食譜，但我再次修改過，看來我也有些進步吧。感謝編輯的認真及對我的瞭解、包容，更感謝背後功臣——攝影師、美術指導，還有我的助手燕姐。更願新知舊雨都能喜愛這本化複雜為簡單的食譜。

祝福大家！

方任利莎

電視節目的開始

　　我是由 1979 年開始主持烹飪節目，從 1982 年起獨自負責整個烹飪節目，叫做《午間小敘》（後於 1988 年改為「方太廣場」），由星期一至星期五，每日介紹一個食譜，即每月至少廿個，每年至少都有三百多個食譜，廿多年來有多少食譜連我自己都記不清了。不過，有些食譜特別受觀眾喜愛，某些食譜又帶來街市購買熱潮，都使人難忘。觀眾的熱情是對我最大的支持和鼓勵，更推動我發掘構思新的食材及美味的食譜，例如：鑊仔叉燒、白米蝦、苔條魚、沙鯭仔等，都是經濟、好味的菜式。

　　我開始做「午間小敘」烹飪主持前，曾和監製有過一次會議，我希望能讓我自主設計食譜、有我自己的風格。我心中的觀眾以家庭婦女為主，所以，我要用心思來做，以經濟、有營養價值、好味、容易學、上枱美觀為目標。很幸運這位監製很通情達理，他不但答應我的請求，並給我很多指導、支持，使我們做紅了這個節目，電視台用最少的投資，換來大的收益。這位當年的監製，就是周偉材先生（現任香港電台助理廣播處長），很感謝他。

　　在「午間小敘」前，是一星期一兩次的烹飪節目，當時是現場直播，雖比較緊張，不容有錯，但只是六、七分鐘的事，但「午間小敘」一個星期有五次，變成一定要預先錄好。一次要準備五個菜式的所有材料，是頗為瑣碎的事，所以，當時我對助手說，只可殺錯，不能放過，即是材料只可買多，不能買少，因為時間寶貴，不容有失。

　　記得有一次是做乳鴿的菜式，因機器故障，要重來一次，當時已十二點，廣播道下面的畔溪酒家也已收工，很是狼狽。結果將乳鴿沖了一個涼，才解決了難題。做節目我從來都是一次過，從不曾 NG，另外每次十二分鐘的烹飪時間，有時快了一些剩下一分鐘，要多講一分鐘話，且要有內容，都不是容易事，不過，一切事肯努力就熟能生巧。我的工作表面看是輕鬆容易，其實，好壞都會在剎那間出現在別人面前，並不是輕鬆和簡單的，多年來我都是在戰戰兢兢中渡過的，是表面輕鬆而已。不過，我還是熱愛我的工作，主要是有觀眾的熱情反應，在熒光幕上建立的感情，使人一生難忘。

　　此外，電視的烹飪工作，是團隊的合作，雖有幕前幕後之分，但各司各職是很重要的，因此大家感情很好，我們雖是同事，但我們習慣叫做手足。這都是我生命中難忘的一頁。

西汁煎雞腿扒

Fried Chicken in
Ketchup Sauce

鮮茄燴雞
Braised Chicken with Tomato

鮮茄燴雞
Braised Chicken with Tomato

材料 Ingredients

冰鮮雞	半隻
薯仔	1 個
番茄	3-4 個
洋葱	1 個
甘筍	半條
乾葱片	少許

1/2 chilled chicken
1 potato
3-4 tomatoes
1 onion
1/2 carrot
sliced shallot

醃料 Marinade

生抽	1 1/2 湯匙
胡椒粉、酒	各少許

1 1/2 tbsp light soy sauce
pepper
wine

調味料 Seasoning

鹽	3/4 茶匙
生抽	2 茶匙
水	約 1 1/3 杯

3/4 tsp salt
2 tsp light soy sauce
1 1/3 cup water

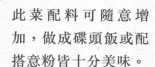

此菜配料可隨意增加，做成碟頭飯或配搭意粉皆十分美味。

做法 Method

1. 雞去頭、腳及尾部，除去腹內肥膏，洗淨斬件，放入醃料拌勻，略醃。
2. 薯仔去皮切塊，略炸；洋葱切塊，略炒；番茄、甘筍切塊。待用。
3. 燒熱油約 1 湯匙，爆炒乾葱，放入雞塊，炒至轉色，盛出放入深鍋中，加入薯仔、番茄、甘筍、調味料同燴煮至材料熟、汁濃。
4. 加入洋葱略燴煮片刻，試味後即成。

1. Remove head, feet, tail and fat from chicken. Rinse and cut into pieces. Mix with marinade.
2. Peel potato and cut into chunks. Deep fry briefly. Cut onion into chunks and stir fry briefly. Cut tomatoes and carrot into chunks. Set aside.
3. Heat 1 tbsp of oil, fry shallot until fragrant and add chicken. Fry until chicken changes colour. Transfer to a deep pot. Add potato, tomato, carrot and seasoning. Simmer until all ingredients cooked and the sauce thickens.
4. Add onion and cook for a while. Taste. Serve.

腰果炒腎球
Fried Turkey Gizzard
with Cashew Nuts

材料 Ingredients

火雞腎　2-3 個
青、紅西椒、洋蔥　各半個
腰果　2-3 両
乾蔥片　少許

2-3 turkey gizzards
1/2 green bell pepper
1/2 red bell pepper
1/2 onion
75-113 g cashew nuts
sliced shallot

醃料 Marinade

酒
胡椒粉 ⎤ 各適量
生粉 ⎦

wine
pepper
caltrop starch

調味料 Seasoning

生抽　2 茶匙
糖　1/4 茶匙
水　1 湯匙

2 tsp light soy sauce
1/4 tsp sugar
1 tbsp water

做法 Method

1. 火雞腎去硬皮，洗淨剕花切成小塊，放入醃料拌勻，飛水後待用。
2. 腰果用熱水泡透後瀝乾，用溫油慢火炸脆，待用。
3. 西椒和洋蔥切成角狀，加少許鹽一同炒熟，待用。
4. 燒熱油約 1 1/2 湯匙，爆香乾蔥，放入腎球，灒酒，加入調味料，再放入西椒、洋蔥等炒勻。
5. 將腰果加入即成。

1. Skin turkey gizzards, rinse, cut cross marks and cut into pieces. Mix with marinade. Blanch.
2. Soak cashew nuts thoroughly with hot water and drain. Deep fry with warm oil over low heat until crispy. Set aside.
3. Cut bell pepper and onion into wedges. Stir fry with some salt. Set aside.
4. Heat 1 1/2 tbsp of oil. Stir fry shallot until fragrant. Add turkey gizzards. Sizzle with wine and add seasoning. Add bell pepper and onion and stir fry well.
5. Stir in cashew nuts. Serve.

腰果用凍油炸較容易控制，用慢火炸至油熱、腰果變黃即成。

蝦 醬 炸 雞 塊
Deep Fried Chicken in Fermented Shrimp Paste

材料 Ingredients

靚光雞　　半隻
蝦醬　　約 3/4 湯匙
蒜茸、乾葱茸　各 1 茶匙
芫茜梗切碎　約 1 湯匙

1/2 dressed chicken
3/4 tbsp fermented shrimp paste
1 tsp grated garlic
1 tsp grated shallot
1 tbsp chopped coriander stalk

醃料 Marinade

酒　　半湯匙
胡椒粉、糖　各少許

1/2 tbsp wine
pepper
sugar

做法 Method

1. 將光雞洗淨斬件，瀝乾水分。
2. 將醃料、蝦醬放入雞塊中拌勻，醃至入味，把雞塊逐件取出，瀝去汁液。
3. 把蒜茸、乾葱茸、芫茜碎加入醃透的雞塊，撲上少許乾粉。
4. 將油燒至大熱，放入雞塊，炸至雞熟及酥脆即可撈起，瀝乾油分上碟。

1. Rinse chicken and cut into pieces. Drain.
2. Mix chicken with marinade and fermented shrimp paste. Let it sit. Take each chicken piece and drain.
3. Mix garlic, shallot and coriander with chicken pieces. Briefly coat with dry caltrop starch.
4. Heat oil thoroughly. Deep fry chicken pieces until crispy. Drain. Serve.

這一道「蝦醬雞」曾經在新加坡風行一時。

銀杏炒鴨舌
Stir Fried Duck Tongue with Ginkgo

材料 Ingredients

急凍鴨舌	12 両
急凍白果	約 3 両
陳皮	1 塊（小）
薑、葱	各少許

450 g frozen duck tongue
113 g frozen ginkgo
1 small piece dried tangerine peel
ginger
spring onion

調味料 Seasoning

生抽	2 茶匙
老抽	半湯匙
糖	1 茶匙
胡椒粉、麻油	各少許
水	半杯

2 tsp light soy sauce
1/2 tbsp dark soy sauce
1 tsp sugar
pepper
sesame oil
1/2 cup water

做法 Method

1. 鴨舌解凍，用薑、葱出水後沖淨，撈起瀝乾，起骨待用。
2. 陳皮用清水浸軟，刮淨內層，切成絲待用。
3. 白果飛水後瀝乾水分，待用。
4. 爆香薑、葱少許，放入鴨舌，潷酒，加入調味料、陳皮略煮片刻。
5. 至汁液收濃時，加入白果同拌炒均勻，即可上碟享用。

1. Defrost duck tongue. Blanch with ginger and spring onion. Rinse, drain and remove bones.
2. Soak dried tangerine peel until soft and remove pith. Shred.
3. Blanch ginkgo and drain.
4. Fry ginger and spring onion until fragrant. Add duck tongue and sizzle with wine. Add seasoning and dried tangerine peel. Cook for a while.
5. When the sauce thickens, add ginkgo and stir fry well. Serve.

急凍鴨舌膻味重，飛水至八成熟、洗淨可去膻味。起骨方法是在鴨舌尾端用手指推上，即可輕易把骨取出。這菜式將鴨舌去骨的原因是容易食，因為可同時享用已去骨的鴨舌和配料，毋須吐骨。若製作滷水鴨舌就不用去骨。

數字雞翼
Chicken Wings in Soy Sauce

數字雞翼
Chicken Wings in Soy Sauce

「數字雞翼」本叫做一二三四五雞翼，主要是調味料中的份量正好是這五個數字，一份酒、二份醋、三份糖、四份豉油（生抽加老抽），五份水。這是一個童子軍露營時的考試食譜，只要小心火候，是一定成功的，且味道好。童子軍成員在成功後，回家向母親顯身手後流傳出來的，可說是溫馨食譜，同時，也是初入廚一定成功的菜餚。當年在電視中介紹，很受歡迎，更令童子軍們十分自豪。

材料 Ingredients

雞翼　6-8 隻
蔥　　3 條
薑片　少許

6-8 chicken wings
3 sprigs spring onion
sliced ginger

調味料 Seasoning

酒　1 湯匙
醋　2 湯匙
糖　3 湯匙
老抽　2 湯匙
生抽　2 湯匙
水　5 湯匙

I tbsp wine
2 tbsp vinegar
3 tbsp sugar
2 tbsp dark soy sauce
2 tbsp light soy sauce
5 tbsp water

做法 Method

1. 雞翼解凍、斬件，洗淨後瀝乾水分，用少許酒拌勻，待用。
2. 燒熱砂鍋仔，放入油約 2 湯匙，爆香薑片、葱段，放入雞翼爆透。
3. 將調味料逐樣加入，用小火燜煮至汁料收乾、雞翼熟透即成。

1. Defrost chicken wings and chop into pieces. Rinse and drain. Mix with some wine. Set aside.
2. Heat a clay pot and add 2 tbsp of oil. Fry ginger and spring onion until fragrant. Add chicken wings and fry.
3. Add seasoning one by one and simmer over low heat until the sauce reduces and chicken wings cooked.

紅糟雞丁
Stir Fried Chicken and Bamboo Shoot in Red Fermented Rice Paste

材料 Ingredients

雞腿　　2 隻
筍肉　　3 両
西芹　　2 枝
紅糟　　1 1/3 湯匙
乾葱　　1 粒（切片）

2 chicken thighs
113 g bamboo shoot
2 stalks celery
1 1/3 tbsp fermented red rice paste
1 clove shallot (sliced)

醃料 Marinade

生抽　　2 茶匙
酒　　　2 茶匙
生粉　　適量

2 tsp light soy sauce
2 tsp wine
caltrop starch

調味料 Seasoning

生抽　　半茶匙
糖　　　半茶匙
酒　　　半湯匙
水　　　約 1/3 杯
麻油　　少許

1/2 tsp light soy sauce
1/2 tsp sugar
1/2 tbsp wine
1/3 cup water
sesame oil

做法 Method

1. 雞腿除皮去骨、筋，切成丁狀，放入醃料拌勻，泡油待用。筍肉煮熟切粒。
2. 西芹斜切成塊，炒熟，先上碟，待用。
3. 燒熱油約 1 湯匙，爆香乾葱片，放入紅糟、調味料煮勻。放入雞丁、筍粒，炒勻，盛起放上西芹面即成。

1. Remove skin, bone and tendon from chicken thighs. Dice and mix with marinade. Deep fry with warm oil until cooked. Boil bamboo shoot and cut into dices.
2. Slice celery at an angle. Stir fry until cooked. Transfer to serving plate.
3. Heat 1 tbsp of oil and fry shallot until fragrant. Add fermented red rice paste and seasoning and mix together. Add chicken and bamboo shoot and stir fry well. Transfer to the serving plate. Serve.

紅糟易焦，須快手炒勻，然後注入調味料。
雞腿肉較嫩滑，但也可用雞胸肉代替。

宗棠雞煲
Braised Chicken in Bamboo Liquor Sauce

　　左宗棠曾任请朝湘軍統帥，勝仗後被封爲大將軍。據說這一味宗棠雞，正是在左宗棠將軍出征在外時，用雞和酒簡單調味煮成的菜式。後經廚師改良用上了左宗棠喜愛的竹葉青酒爲主要調味料。我試過有店家將雞起肉，配合辣椒炒成，也稱之爲左宗棠雞，但我想在行軍時不太可能做出如此精緻的菜式。我的宗棠雞煲，學自湖南老廚師，原汁原味且易煮，不妨一試。

材料 Ingredients

光雞　半隻
冬菇　4-6 隻
薑　1 塊 (約 4 両)
葱　6 條

1/2 dressed chicken
4-6 dried shiitake mushrooms
150 g ginger
6 sprigs spring onion

調味料 Seasoning

竹葉青酒　4 湯匙
生抽　2 湯匙
糖　3/4 湯匙
鹽　1/4 茶匙
老抽　1 茶匙
麻油、胡椒粉　各少許

4 tbsp Chu Yeh Ching liquor
2 tbsp light soy sauce
3/4 tbsp sugar
1/4 tsp salt
1 tsp dark soy sauce
sesame oil
pepper

做法 Method

1. 雞洗淨斬件，冬菇浸軟去蒂，切件，薑去皮切片，葱切段。
2. 將瓦煲仔燒熱放入油約 1 湯匙，放入半份薑片、葱段爆香。
3. 將雞塊放入略爆香，將所有調味料混合加入拌勻，蓋上蓋用中火煮至滾起。
4. 改用小火燜至雞熟，放入餘下半份薑和葱，再焗煮約 3 分鐘，即可原煲上枱享用。

1. Rinse chicken and chop into pieces. Soak dried shiitake mushrooms until soft, remove stalks and cut into pieces. Peel ginger and slice. Section spring onion.
2. Heat a clay pot and add 1 tbsp of oil. Fry half of the ginger and spring onion until fragrant.
3. Add chicken and fry until fragrant. Mix together the seasoning and stir into the clay pot. Cover the pot and bring to the boil over medium heat.
4. Turn to low heat and simmer until chicken cooked. Add remaining ginger and spring onion. Cover and cook for 3 minutes. Serve.

薑片可多些，如用子薑會更美味。雞含有水分，故毋須加水，原汁原味是這道菜的特色。

疑幻似真的菜名

中菜的命名和西菜有很大的分別，在西菜的菜牌上常會把所用的材料列出，而中菜的命名卻各有不同，好像富貴雞本叫做叫化雞，「叫化」即是普通話中的乞丐（叫化子）。據說，明末清初，江蘇省常熟縣，有一乞丐偷得一雞，殺後無炊具，連毛塗上泥巴，放入火中烤，泥乾敲去泥殼，香氣四溢，乞丐大喜就狼吞虎嚥的吃了起來。恰巧明朝大學士錢謙益經過，聞到香味，便命隨從查詢，回家後命家廚調味並加入改良，更覺鮮美。因為乞丐名字不雅，後改稱為富貴雞。

中國的飲食，有時有趣，有時也莫名其妙，揚州本地並無揚州炒飯、新加坡也沒有星州炒米，想來一定是首創者吸引食客的一種手法。其實最重要的是好吃。

酒家做富貴雞，以前是利用酒罈上面的泥敲碎混合水再用，現在大多數是用塘泥了，我感覺最老土是在上枱時叫主家敲碎泥土的儀式，我就會叫小孩去做這儀式玩玩了。在家中做，實在不要找這種麻煩，我們用簡化改良的方法，同樣可保原汁原味。

在食材方面也同樣有趣，例如蝴蝶腩，名字很美，其實是豬腹中的肉，是附在內臟豬肺附近的軟肉，本來是多餘的物資，大約五十年代香港屠房還設在土瓜灣時，這些蝴蝶腩都是屠房工作人員的下欄，在街邊賣約八毫子一斤，節儉的主婦們買回家煲湯用，不能上大枱的。真叫做十八年風水輪流轉，現在要在肉枱預定才有，雖然不是貴，但多數交給酒家，現代年輕的一代沒挨過窮，他們喜歡新鮮，加上美麗的名字「蝴蝶腩」，真新奇！下價的食材都變貴了。

中國菜也有幻想的菜式，好像魚香茄子、魚香肉絲根本一點魚腥都沒有，只因為魚香的菜式來自四川；四川是盆地，當年交通不便，所以沒有魚，但在中國的習俗，喜慶宴席不能無魚，魚代表富貴有餘，因此就想出用一種煮魚的汁料，將木製的魚放上碟，淋上煮魚的汁，上枱，大家也明白只是取好兆頭，所以，吃些汁就代表圓滿了。此種汁給人的味覺是有魚的香味，很惹味，因此後來就被運用在其他材料中，成為魚香茄子、魚香肉絲，更可做成魚香豆腐。雖說簡單，但要用到恰到好處並不容易，就要靠多觀察、多嘗試及對材料有深切的認識，配合得恰當，才能煮出好味的菜餚。看似最簡單，有時會是最複雜。就靠領會，也可說是要悟出個道理吧！

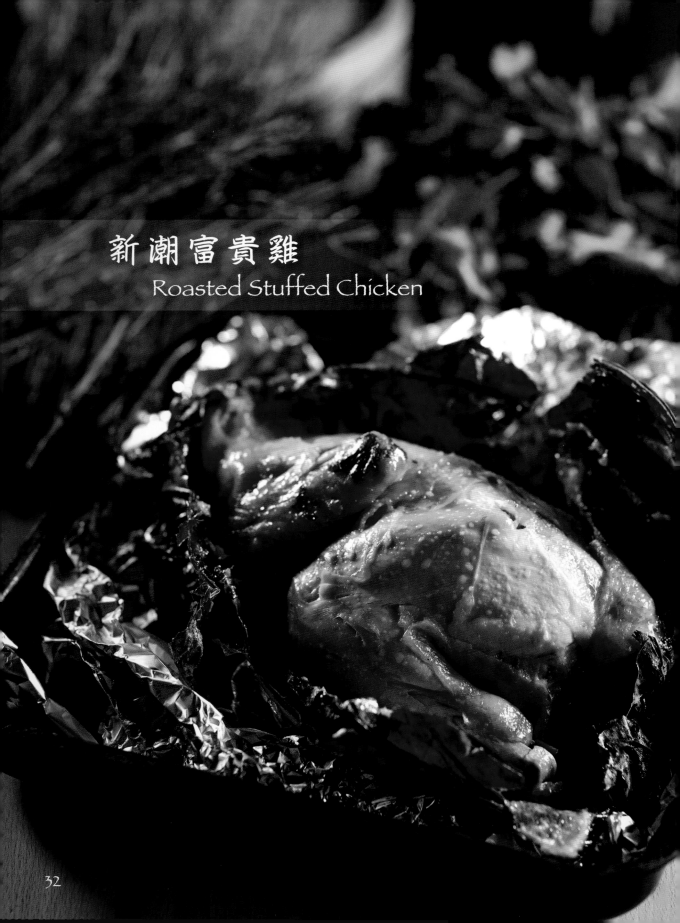

新潮富貴雞
Roasted Stuffed Chicken

材料 Ingredients

光雞　1 隻
半肥瘦豬肉　4 両
冬菇　4 朵
雲南大頭菜　1 塊（小）
金針、雲耳　各適量
葱段、薑絲、乾葱片　各少許
乾荷葉　1 張

1 dressed chicken
150 g half-fat pork
4 dried shiitake mushrooms
1 small piece Yunnan preserved turnip
dried lily buds
black fungus
spring onion sections
shredded ginger
sliced shallot
1 sheet dry lotus leaf

醃料 Marinade

生抽　1 1/2 茶匙
生粉　1 茶匙
水　1 湯匙

1 1/2 tsp light soy sauce
1 tsp caltrop starch
1 tbsp water

調味料 Seasoning

生抽　1 1/2 茶匙
糖　1 茶匙
水　約 1/3 杯

1 1/2 tsp light soy sauce
1 tsp sugar
1/3 cup water

新潮富貴雞
Roasted Stuffed Chicken

做法 Method

1. 雞洗淨，用酒、胡椒粉略醃，待用。冬菇浸透切絲。
2. 豬肉、大頭菜切絲，金針、雲耳浸透洗淨。肉絲放入醃料。
3. 燒熱油約 1 1/2 湯匙，放入豬肉絲炒熟，放入大頭菜、金針、雲耳、冬菇，加入調味料煮勻，並加入葱段、薑絲、乾葱片，炒勻，盛出待用。
4. 將上項材料釀入雞肚中，用荷葉包住，再包上錫紙，放入焗爐中，用 250°C 火焗約 1 小時即成。

荷葉拖水取出後，搽油才用。

1. Rinse chicken. Marinate with wine and pepper. Set aside. Soak dried shiitake mushrooms until soft and shred.

2. Shred pork and preserved turnip. Soak dried lily buds and black fungus thoroughly and rinse. Mix pork with marinade.

3. Heat 1 1/2 tbsp of oil. Stir fry pork until cooked. Add preserved turnip, dried lily buds, black fungus and shiitake mushrooms. Add seasoning and mix well. Add spring onion, ginger and shallot. Stir fry well to finish the filling. Set aside.

4. Stuff the filling inside the chicken and wrap the chicken with lotus leaf. Wrap with aluminium foil. Put in an oven and bake at 250°C for 1 hour. Serve.

蘿蔔蝴蝶腩
Braised Pork with White Radish

水晶皮凍
Chilled Pork Skin Aspic

水晶皮凍
Chilled Pork Skin Aspic

材料 Ingredients

新鮮豬皮	1 塊（約 6 両）
八角	1 粒
雞精	1 粒
芫茜	1 棵
甘筍茸	約 2 湯匙
水	約 2 1/2 杯
魚膠粉	約 1 1/2 湯匙

1 fresh pork skin (about 225 g)

1 star anise

1 chicken bouillon cube

1 sprig coriander

2 tbsp chopped carrot

2 1/2 cups water

1 1/2 tbsp gelatin powder

調味料 Seasoning

老抽	半湯匙
生抽	1 1/2 茶匙
糖	半茶匙
胡椒粉	少許

1/2 tbsp dark soy sauce

1 1/2 tsp light soy sauce

1/2 tsp sugar

pepper

豬皮價廉，甚至某些相熟肉檔會不
收錢，此食譜含豐富膠質，有美容
功效，但必須切淨肥脂。

做法 Method

1. 豬皮刮淨豬毛及皮下肥脂，洗淨飛水。
2. 將豬皮加入八角、雞精、水，煮至豬皮腍及水的份量剩下約 1 1/2 杯，取出豬皮，
   切成小粒後放回，再放入調味料拌勻同煮，並將八角撈出。待用。
3. 將魚膠粉用約 3 湯匙溫水調勻，注入上項材料中拌勻，並加入切碎的芫茜、甘筍茸，
   拌勻，盛出放入深盤中，放入雪櫃，雪至凝固即可切件上碟。

1. Peel off any hair and fat from pork skin. Rinse and blanch.
2. Cook together pork skin, star anise, chicken bouillon and water until pork skin softens
   and sauce reduces to 1 1/2 cups. Cut pork skin into small dices and put back into the
   pot. Mix in seasoning and cook for a while. Discard star anise. Set aside.
3. Mix gelatin powder with 3 tbsp of warm water. Mix into the above mixture and add
   chopped coriander and carrot. Transfer the mixture into a deep tray and chill until
   set. Cut into pieces and serve.

沙茶螞蟻上樹
Fried Mung Bean Vermicelli in
Satay Sauce

材料 Ingredients

粉絲	1 両
剁碎豬肉	3 両
洋蔥粒	半杯
蒜茸	1 茶匙
青椒粒	
甘筍粒	各 1/4 杯
芹菜粒	
沙茶醬	1 1/2 湯匙

38 g mung bean vermicelli

113 g chopped pork

1/2 cup chopped onion

1 tsp grated garlic

1/4 cup diced green bell pepper

1/4 cup diced carrot

1/4 cup diced celery

1 1/2 tbsp satay paste

調味料 Seasoning

生抽	半湯匙
糖	1/4 茶匙
麻油、胡椒粉	各少許
水	3 湯匙

1/2 tbsp light soy sauce

1/4 tsp sugar

sesame oil

pepper

3 tbsp water

做法 Method

1. 粉絲浸透，瀝乾水分，待用。豬肉放入生抽 2 茶匙、生粉、胡椒粉各少許，拌勻待用。
2. 燒熱油約 2 湯匙，放入洋葱粒、蒜茸、豬肉碎同炒勻至熟。
3. 將青椒粒、甘筍粒加入並放入沙茶醬炒勻，加入粉絲、芹菜粒、調味料，炒拌均勻即可上碟享用。

1. Soak mung bean vermicelli thoroughly and drain. Mix pork with 2 tsp of light soy sauce, some caltrop starch and pepper.
2. Heat 2 tbsp of oil. Stir in onion, grated garlic and pork. Stir fry until cooked.
3. Add bell pepper, carrot and satay paste. Stir fry well and add mung bean vermicelli, celery and seasoning. Serve.

想再簡單點可減少配料，只用碎肉、粉絲、洋葱，便已十分惹味。

紅燒五香肋排
Suzhou-style Braised Pork Rib

紅燒五香肋排
Suzhou-style Braised Pork Rib

材料 Ingredients

肋排	約 1 斤
葱	2 條
薑	2 片
酒	少許
八角	2 粒
水	約 1 1/2 杯
蜂蜜	約 1 湯匙

600 g pork ribs

2 sprigs spring onion

2 ginger slices

wine

2 star anises

1 1/2 cups water

1 tbsp honey

調味料 Seasoning

老抽	約 3 湯匙
鹽	1/3 茶匙
冰糖碎	2 湯匙

3 tbsp dark soy sauce

1/3 tsp salt

2 tbsp crushed rock sugar

材料 Ingredients

新鮮大樹菠蘿	約 6 件
瘦肉	6 両
椰汁	半杯
蒜茸、乾葱茸	各半湯匙
紅椒圈	1 湯匙
蝦醬	2 茶匙
薑米	少許

6 pieces fresh jackfruit
225 g lean pork
1/2 cup coconut juice
1/2 tbsp grated garlic
1/2 tbsp grated shallot
1 tbsp red chilli rounds
2 tsp fermented shrimp paste
finely chopped ginger

調味料 Seasoning

鹽	1/4 茶匙
糖	1/3 茶匙

1/4 tsp salt
1/3 tsp sugar

做法 Method

1. 將大樹菠蘿去核切件。瘦肉切片，放入酒、生粉各少許拌勻，待用。
2. 燒熱油約 2 湯匙，爆香蒜、乾葱和薑，放入肉片炒至半熟，加入蝦醬、紅椒炒勻，注入椰汁及少許水分至豬肉熟透。下調味料，放入大樹菠蘿即成。

1. Seed jackfruit and cut into pieces. Slice lean pork and mix in wine and caltrop starch. Set aside.
2. Heat 2 tbsp of oil and fry garlic, shallot and ginger. Add lean pork and stir fry until half cooked. Mix in fermented shrimp paste and red chilli. Pour in coconut juice and water. Cook until pork fully cooked. Stir in seasoning and jackfruit. Serve.

大樹菠蘿又叫做菠蘿蜜，如無新鮮的可選用罐裝，但須瀝去糖水。此菜式為娘惹菜。

在新加坡的日子

在新加坡電視台工作時，最開心是在錄影後，晚上和導演及我的助手一起去晚飯，我們四處找特別的食品共享。我的助手常說忙了一整天，難道不累嗎，還要去玩？正因為忙了一整天，除了體力之外，實在精神很緊張，要面對和應付的事太多了，真的需要放鬆一下，所以有時我們會去唱歌，這都是助手不明白的。

我們很喜歡光顧一間家庭式的小飯館，店名都不記得了，因為我們稱店主為「家嫂」，一說去「家嫂」處大家就明白了，地址近唐人街，由兩代人打理，媳婦和家公是主廚，沒有菜牌，但食物非常好吃，例如將墨魚仔炸脆，沾上蜜汁灑上芝麻，我們總是菜沒來已吃光兩盤。總之是有甚麼就吃甚麼，完全是店東作主。因為有很多同事都光顧，大家很熟悉，好像回家吃飯一般。導演是馬來西亞人，我是香港去的，所以在一天辛勞後去「家嫂」處吃飯，有另一種溫暖。在「家嫂」處吃了蝦醬炸雞塊、菠蘿蜜煮肉片、白花釀油條、胡椒肉骨等好味的菜，我忍不住偷師，在香港的電視節目中介紹給觀眾，都受到大家的喜愛，這就是分享的快樂了。

家嫂很歡迎我和導演，主要是我們每次都叫很多菜，吃不完就算了。因為我想學習，她感覺我很闊氣，不計較、捨得花錢，所以待我們很好。可是，我們也有讓她生氣的時候。試過我們坐下後，點了菜忽然想起，約了人在別處等，就立即說不要煮了，我們忘了約了人。有時，突然改變主意想吃日本菜了，我們又會離開了。當我們向家嫂道歉時，她就大聲說：「少來這套了，我都受慣你們這種作風了。」我們也不管她生氣就走了。如今想起，仍回味無窮。

我和這位導演共同工作約十年，他的年紀和我兒子差不多，我們是最佳拍檔、知己老友，又有母子般的感情，這都是我人生路上可遇不可求的收穫，我十分珍惜。可惜，這位導演在四年前去了天國，使我十分傷感。在我去馬來西亞拜祭他時，看着他的相片，忽然想起他說過的話：一切的事，試過、經歷過就是最好的了。我們在一起工作時，我們拼命要做到最好，但我們也拼命的玩樂。有一次我們為了想去泰國玩，把四天的工作在兩天做完。結果到曼谷後在酒店睡了兩天，第三天在咖啡室見面，他問我新加坡沒有床嗎？我們來曼谷睡覺？結果，我們在曼谷吃了一頓飯，就趕回新加坡了。這種瀟灑快樂只有我和導演占美才明白、才會做。

鎮江排骨
Pork Ribs in Zhenjiang Vinegar

鎮江排骨
Pork Ribs in Zhenjiang Vinegar

材料 Ingredients

肉排　12 両
薑　2 片
葱　2 條（切段）
紅糟　1 茶匙（先爆炒）

450 g spare ribs
2 ginger slices
2 sprigs spring onion (sectioned)
1 tsp fermented red rice (stir fried)

醃料 Marinade

鹽　1/4 茶匙
生抽　1 1/2 湯匙
酒　1 湯匙
胡椒粉　少許

1/4 tsp salt
1 1/2 tbsp light soy sauce
1 tbsp wine
pepper

調味料 Seasoning

鎮江香醋　6 湯匙
冰糖　57 克
鹽　1/4 茶匙
水　1/3 杯

6 tbsp Zhenjiang vinegar
57 g rock sugar
1/4 tsp salt
1/3 cup water

鎮江排骨做法有多種，正所謂各師各法，但主要用鎮江出產的醋。加入紅糟少許，可增色澤；但紅糟是福建名產。據說這是老師父的改良之作，是否加入可隨意。但此菜一定要味道香濃才夠味、夠格。

做法 Method

1. 將排骨斬件，放入醃料內醃約半小時，再加入粟粉 1 湯匙，拌勻待用。
2. 用半鑊燒至大熱的油，慢火將排骨炸至金黃色，撈起待用。
3. 薑、葱爆香，加入排骨、紅糟炒勻，潷酒，盛起。
4. 調味料煮滾，加入排骨煮至汁略收乾，加入麻油 1 茶匙即成。

1. Cut spare ribs into pieces. Mix with marinade and let it sit for 30 minutes. Mix in 1 tbsp of cornstarch.
2. Heat half wok of oil. Deep fry spare rib over low heat until golden brown. Set aside.
3. Fry ginger and spring onion until fragrant. Stir in spare rib and fermented red rice. Sizzle with wine. Remove.
4. Bring the seasoning to boil and put in spare rib. Cook until the sauce reduces. Add 1 tsp of sesame oil. Serve.

材料 Ingredients

豬脷	1 條
洋葱	
薯仔	各 1 個
甘筍	
乾葱、蒜肉	各 1 粒
咖喱粉	約 1 1/2 湯匙
淡奶	少許

1 pork tongue
1 onion
1 potato
1 carrot
1 clove shallot
1 clove garlic
1 1/2 tbsp curry powder
evaporated milk

在一片物價高漲聲中，豬脷是價廉物美的食材，只是需要做些事前潔淨工作，先要略煮把脷苔刮至十分乾淨。豬脷可先用來煮湯，再撈出煮咖喱，一味兩用，經濟可口。

做法 Method

1. 豬脷放入清水中煮至外層呈白色，取出刮去外層，洗淨。放入湯中煮至熟透，取出切成小片，待用。
2. 洋葱切塊，甘筍、薯仔去皮切塊，乾葱、蒜肉同剁碎。
3. 燒熱油約 1 湯匙，爆香料頭放入豬脷片、咖喱粉同炒勻，注入適量水分（以蓋過材料面為準）。
4. 將洋葱、甘筍、薯仔同加入上項材料中，燴煮片刻，加入約 1 茶匙鹽，最後加入少許淡奶即成。

1. Boil pork tongue until the surface whitened. Remove and scrape off the surface. Rinse. Boil in soup until thoroughly cooked. Slice.
2. Cut onion into pieces. Peel carrot and potato and cut into pieces. Chop shallot and garlic.
3. Heat 1 tbsp of oil. Fry shallot and garlic until fragrant. Stir in pork tongue and curry powder. Pour in water until covered.
4. Add onion, carrot and potato and cook for a while. Add 1 tsp of salt. Pour in evaporated milk. Serve.

咖喱燜豬脷
Stewed Pork Tongue in Curry Sauce

黑椒牛仔骨
Stir Fried Beef Short Ribs

材料 Ingredients

牛腱　約 6 両
韮菜花　6 両
蜆芥醬　約 2 湯匙
薑米　少許

225 g beef shank
225 g garlic chive
2 tbsp fermented clam sauce
finely chopped ginger

醃料 Marinade

生抽　2 茶匙
胡椒粉、生粉　各少許
油　半湯匙

2 tsp light soy sauce
pepper
caltrop starch
1/2 tbsp oil

調味料 Seasoning

糖　1/4 茶匙
麻油　少許

1/4 tsp sugar
sesame oil

做法 Method

1. 韮菜花切去尾段過老部分，洗淨切段，炒熟上碟，待用。
2. 牛腱切薄片，放入醃料拌勻。
3. 燒鑊至極熱，放油約 1 1/2 湯匙爆香薑米，將牛腱爆炒至熟，放入蜆芥醬、調味料快手炒勻，盛出放上韮菜花面即成。

1. Cut off the mature part from the end of garlic chives. Rinse and section. Stir fry and set aside on plate.
2. Slice beef shank and mix in marinade.
3. Heat wok thoroughly. Add 1 1/2 tbsp of oil to fry ginger until fragrant. Stir fry beef shank until cooked. Add fermented clam sauce and seasoning and stir fry quickly. Transfer on top of garlic chives on the plate. Serve.

蜆芥不宜煮得太久，要快手；它的味道偏鹹，須留意。

香橙百寶肉丁

Stir Fried Pork with Orange Zest and Vegetables

材料 Ingredients

瘦肉	4 両
豆腐干	4 塊
筍肉	約 3 両
冬菇	6 朵
青、紅西椒	各 1 個
蝦米	2 湯匙
蒜茸	2 茶匙
磨豉醬	約 3 湯匙
橙皮	1 個（除去白色果衣，浸熱水片刻取出，切粒）

150 g lean pork

4 pieces dried beancurd

113 g bamboo shoot

6 dried shiitake mushrooms

1 green bell pepper

1 red bell pepper

2 tbsp dried shrimps

2 tsp grated garlic

3 tbsp ground bean paste

1 orange zest (remove pith, soak with hot water and chop)

調味料 Seasoning

糖	約 1 3/4 湯匙
桂林辣椒醬	1 1/2 茶匙
茄汁	2 茶匙
麻油	少許
水	約 1/3 杯

1 3/4 tbsp sugar

1 1/2 tsp Guilin chilli sauce

2 tsp ketchup

sesame oil

1/3 cup water

做法 Method

1. 豆腐干、筍肉與浸透的冬菇同切成粒狀。青、紅椒去籽洗淨，切粒。蝦米沖水瀝乾。
2. 瘦肉切成粒狀，放入生抽2茶匙及生粉少許拌勻，用少許油炒熟，待用。
3. 燒熱油約1 1/2湯匙，放入蒜茸、肉丁、蝦米、冬菇、豆腐干、筍肉、青、紅椒炒勻，再加入磨豉醬及橙皮粒同炒，煮片刻。
4. 加入調味料，可試味後隨個人口味，略為加減味道。

1. Cut dried beancurd, bamboo shoot and soaked shiitake mushrooms into small dices. Remove seeds from bell peppers, rinse and cut into dices. Rinse dried shrimps and drain.
2. Cut lean pork into dices. Mix in 2 tsp of light soy sauce and caltrop starch. Stir fry with oil until cooked. Set aside.
3. Heat 1 1/2 tbsp of oil. Add grated garlic, pork, dried shrimps, shiitake mushrooms, dried beancurd, bamboo shoots and bell peppers and stir fry well. Add ground bean paste and orange zest. Cook for a while.
4. Add seasoning to suit your taste. Serve.

筍肉用罐裝的較方便，新鮮的筍必須先煮熟才能用。加入橙皮除了較有特色，更令菜式帶有清新的橙皮香味。

胡椒肉骨
Braised White Pepper Pork Rib

材料 Ingredients

腩排　12 兩
乾蔥茸、蒜茸　各 2 茶匙
紅椒粉、胡椒粉　各約 1 茶匙
白胡椒粒　半茶匙
蔥粒　半湯匙

450 g spare ribs
2 tsp grated shallot
2 tsp grated garlic
1 tsp paprika
1 tsp pepper
1/2 tsp white peppercorns
1/2 tbsp chopped spring onion

醃料 Marinade

生抽　1 1/2 湯匙
糖
胡椒粉 ⎫ 各半茶匙
紅椒粉 ⎭
酒　半湯匙

1 1/2 tbsp light soy sauce
1/2 tsp sugar
1/2 tsp pepper
1/2 tsp paprika
1/2 tbsp wine

調味料 Seasoning

黑甜醬油　1 湯匙
魚露　2 湯匙
片糖　半塊
鹽　少許

1 tbsp black soy sauce
2 tbsp fish sauce
1/2 raw slab sugar
salt

　　排骨是很普通的食材，分肋排、腩排、肉排和腓骨，其實就是分有些適合燜，有些適合炸和蒸，在購買時向肉枱負責人說出烹調方法，他們都會清楚介紹購買哪一類的排骨較為適合。

　　我們在這本書，向大家介紹三款用排骨做的菜餚，雖然同樣是排骨，但用不同方法不同調味，就變成各有不同口味的排骨。其實和人生有些相似，在我們的成長過程中，有些像烹調，你肯努力用心思去烹調，你的人生一樣會豐富、美味，且色、香、味俱全。

做法 Method

1. 將排骨洗淨，抹乾水分，斬成 2 吋餘長段，放入醃料拌勻，放置約 1 小時，略炸撈起，待用。
2. 爆香乾蔥、蒜茸，加入胡椒粉、紅椒粉，再放入走油後的排骨拌炒均勻，灒酒少許。
3. 將上項材料轉放入煲仔中，加入調味料、白胡椒粒及過面的水分，煮滾後，用小火燜煮至肉骨睩，汁料濃帶有黏性，灑入蔥粒，即可上碟享用。

1. Rinse spare ribs and wipe dry. Chop into 2 inch sections. Mix in marinade and let it sit for 1 hour. Deep fry briefly and set aside.
2. Fry shallot and garlic until fragrant. Mix in pepper and paprika. Add spare rib and stir fry well. Sizzle with wine.
3. Transfer to a small pot. Add seasoning, white peppercorns and water to cover all ingredients. Bring to the boil, turn to low heat and simmer until spare rib softened and the sauce is sticky. Sprinkle with spring onion. Serve.

用胡椒粒燜排骨，帶辣、惹味，更有驅濕功效。

鹹魚薯球
Deep Fried Potato Ball with Salted Fish

材料 Ingredients

桂花魚	1 條
薑片、葱段	各少許
紹酒、鹽、胡椒粉	各適量
五柳菜	1 份
薑絲、乾葱片	各少許

1 mandarin fish
sliced ginger
spring onion shreds
Shaoxing wine
salt
pepper
1 serving five-pickled vegetables
shredded ginger
sliced shallot

調味料 Seasoning

浙醋	3 湯匙
糖	約 2 茶匙
鹽	1/4 茶匙
生抽	1 茶匙
水	約 1/3 杯
茄汁	1 湯匙（隨意選用）

3 tbsp Zhejiang vinegar
2 tsp sugar
1/4 tsp salt
1 tsp light soy sauce
1/3 cup water
1 tbsp ketchup (optional)

做法 Method

1. 魚去鱗劏肚洗淨，用少許紹酒、鹽、胡椒粉略醃，再抹上生粉少許待用。
2. 五柳菜全部切絲，待用。
3. 燒滾大半鑊水，放入薑片、葱段、少許紹酒及鹽，將魚放入，熄火，將魚浸約 15 分鐘，再次開火至水再度滾起，將魚撈起，上碟待用。
4. 燒熱油約 1 湯匙，爆香薑絲、乾葱片，放入五柳菜、調味料，煮勻，用少許生粉水埋薄芡，淋上魚身即成。

1. Scale and gut mandarin fish. Marinate with Shaoxing wine, salt and pepper. Briefly coat with caltrop starch.
2. Shred five-pickled vegetables. Set aside.
3. Add water to wok till it is 3/4 full and bring to the boil. Add ginger, spring onion, Shaoxing wine and salt. Add fish and turn off heat. Let fish cook for 15 minutes. Turn on heat and bring to the boil again. Transfer fish to serving plate.
4. Heat 1 tbsp of oil and fry shredded ginger and sliced shallot until fragrant. Add five-pickled vegetables and seasoning and cook for a while. Thicken with caltrop starch mixture. Pour the sauce over fish. Serve.

如用茄汁可增鮮艷色彩，使菜式更為色香味俱全，但可隨意。除了用整條魚，也可改用鯇魚腩。

百花釀油條
Stuffed Chinese Fried Sticks

材料 Ingredients

蝦仁　約半斤
馬蹄　4-6 粒
油條　2 條
乾葱　2 粒（切片）

300 g shelled shrimps
4-6 water chestnuts
2 Chinese fried sticks
2 cloves shallot (sliced)

調味料 Seasoning

茄汁　2 湯匙
喼汁　半湯匙
糖　　2 茶匙
水　　2 湯匙

2 tbsp ketchup
1/2 tbsp Worcestershire sauce
2 tsp sugar
2 tbsp water

可以用平底易潔鑊煎，代替油炸。

做法 Method

1. 蝦仁挑腸洗淨，吸乾水分拍爛，打成蝦膠。

2. 將馬蹄去皮剁碎加入蝦膠中，並放入鹽、胡椒粉、生粉各少許，攪拌均勻，略雪片刻待用。

3. 將每條油條撕開成兩條，切成吋餘長段，剦開中央少許，釀入蝦膠。

4. 將上項材料放入七成熱油中炸至蝦膠熟，撈起瀝乾油分待用。

5. 爆香乾葱片，放入調味料煮勻，盛起，伴百花釀油條享用。

1. Devein and rinse shrimps. Wipe dry and mash into minced shrimp.

2. Peel and chop water chestnuts. Mix into minced shrimp. Add salt, pepper and caltrop starch. Mix together well and refrigerate.

3. Tear each Chinese fried sticks into two and cut into long sections. Cut open each piece and stuff minced shrimp inside.

4. Deep fry stuffed Chinese fried sticks in hot oil until the minced shrimp is cooked. Drain.

5. Fry shallot until fragrant and mix in seasoning. Remove and serve with Stuffed Chinese Fried Sticks.

鍋塌蟹柳
Braised Crab Sticks

材 料 Ingredients

蟹柳	8-10 條
葱粒	半湯匙
薑粒	半茶匙
乾葱粒	半湯匙

8-10 crab sticks
1/2 tbsp chopped spring onion
1/2 tsp chopped ginger
1/2 tbsp chopped shallot

麵 漿 材 料 Batter Ingredients

麵粉	約 4 湯匙
雞蛋	1 個
水	1/4 杯

4 tbsp flour
1 egg
1/4 cup water

調 味 料 Seasoning

生抽	半湯匙
鎮江醋	1 1/2 湯匙
糖	2 茶匙
水	1 湯匙
麻油	少許

1/2 tbsp light soy sauce
1 1/2 tbsp Zhenjiang vinegar
2 tsp sugar
1 tbsp water
sesame oil

做法 Method

1. 蟹柳解凍後，撲上乾生粉少許。
2. 將麵漿材料拌勻成漿狀，將蟹柳沾上麵漿。
3. 在平底鑊放入少許油，逐條放入沾上麵漿的蟹柳，使用小火煎至表面呈微黃、麵漿熟。
4. 將蔥粒、薑粒、乾蔥粒加入並略轉動，再注入調味料，煮至材料吸收汁料，即可上碟享用。

1. Defrost crab sticks and briefly coat with caltrop starch.
2. Mix the batter ingredients and coat crab sticks.
3. Grease non-sticky pan and arrange crab sticks on it. Fry over low heat until the batter slightly browned and cooked.
4. Add spring onion, ginger and shallot and toss lightly. Add seasoning and cook until the sauce dries. Serve.

這是一道簡單易做、受孩子喜愛的菜式。

醬炒毛蟹
Stir Fried Crab with Soybeans

材料 Ingredients

小湖蟹	4-5 隻
新鮮毛豆粒	約 2 両
薑片、蔥段	各適量
乾蔥片	少許
磨豉醬	約 3/4 湯匙
芫茜	1 棵

4-5 small lake crabs

75 g fresh young soybeans

sliced ginger

spring onion sections

sliced shallot

3/4 tbsp ground bean paste

1 sprig coriander

調味料 Seasoning

生抽	半湯匙
糖	2 茶匙
胡椒粉、麻油	各少許
水	約 3/4 杯
酒	約 1 湯匙

1/2 tbsp light soy sauce

2 tsp sugar

pepper

sesame oil

3/4 cup water

1 tbsp wine

· 豆腐可不炸，切成小塊放入同煮。
· 白米蝦即是用未長大的河蝦仔（蝦毛）醃後曬乾而成。上海南貨店有售。

材料 Ingredients

白米蝦　　約 2 湯匙
豆腐　　　1 塊（大）
葱粒　　　半湯匙
芫茜　　　少許
蒜肉　　　1 粒

2 tbsp dried baby shrimps
1 large cube beancurd
1/2 tbsp chopped spring onion
coriander
1 clove garlic

調味料 Seasoning

生抽　　1 湯匙
鹽　　1/3 茶匙
糖　　半茶匙
水　　半杯
麻油　　少許

1 tbsp light soy sauce
1/3 tsp salt
1/2 tsp sugar
1/2 cup water
sesame oil

做法 Method

1. 豆腐切粒，瀝去水分，煎至微黃色（或用熱油炸至呈金黃色）後，瀝去油分，待用。
2. 燒熱油約 1 湯匙，放入拍碎的蒜肉，將豆腐回鑊，放入白米蝦，灒酒。
3. 將調味料加入煮至滾起，使豆腐入味。
4. 放入葱粒、切碎的芫茜，用少許生粉水埋芡，即可趁熱上碟享用。

1. Dice beancurd and drain. Fry until slightly browned (or deep fried until golden brown) and drain.
2. Heat 1 tbsp of oil, add crushed garlic, beancurd and dried baby shrimps. Sizzle with wine.
3. Add seasoning and bring to the boil.
4. Add chopped spring onion and coriander and thicken the sauce with caltrop starch mixture. Serve hot.

街市熱潮

在我的烹飪節目中，我喜歡介紹一些被人忽視或認為不能上大枱的食材，主要是便宜的。記得我曾介紹過一款上海家常小菜「白米蝦煮豆腐」，白米蝦即是未長大的小河蝦，也是廣府人說的「蝦毛」，蝦殼還沒硬很細小，漁民網到時是無法出售的，但棄之可惜，就用少許鹽醃，醃乾連殼像小蝦米般。江南一帶的主婦們用來煮豆腐，加入蛋花湯中，或雲吞、豆腐花中，十分鮮味可口。本是廢物利用，但蝦殼含有鈣質，真可說是中國漁民無意中產生的智慧。

當年我將「白米蝦煮豆腐」的食譜在電視上介紹給觀眾，並告訴大家如有興趣可去九龍城三陽南貨店買一兩元白米蝦，可分數次用。因為我的節目是預錄的，所以根本忘了是幾時播出，有一天經過三陽，老闆叫住我，並大聲對我說：「方太請你下次介紹貴一點的材料，這幾天很多人來買一兩元的白米蝦，看似旺市，但只是一兩元的生意，真是多得你不少了。」我說，有人幫襯是好事啊，難道你要個個人都來買整隻的金華火腿嗎？

沙鯭仔就是小條的剝皮魚，那年頭代剝皮賣六元一斤、自剝皮賣四元一斤。因是賤價魚，潮州人喜歡，廣府人少吃。我用小沙鯭做成椒鹽沙鯭，和煎煮沙鯭的菜式，想不到大受觀眾的喜愛，但是，沙鯭仔卻由幾元變成十元至十二元一斤了。我曾向魚枱抗議，老闆卻笑着說：「可以便宜些賣給你。但方太你可知道一早很多人要我們代剝皮，別的生意都被阻住了。」真是無奈。他們更說：「上次你介紹釀鯪魚，西人太太也來湊熱鬧，請你下次介紹貴價些的魚類。」

這些都是老街坊的有趣人情味舊事，相信當年的主婦們一定還記得這些事吧。太多人對我說看我節目長大，吃媽媽學我煮的菜，好味道。真感謝大家對我的關愛。我們一起經歷過這些歲月，使人難忘。

百花平安卷
Deep Fried Shrimp Stick with Apple

百花平安卷
Deep Fried Shrimp Stick with Apple

材料 Ingredients

蘋果	2 個
蝦膠	約 6 両
泰式雞醬	適量

2 apples
225 g minced shrimp
Thai chicken sauce

調味料 Seasoning

鹽	1/3 茶匙
胡椒粉、生粉	適量

1/3 tsp salt
pepper
caltrop starch

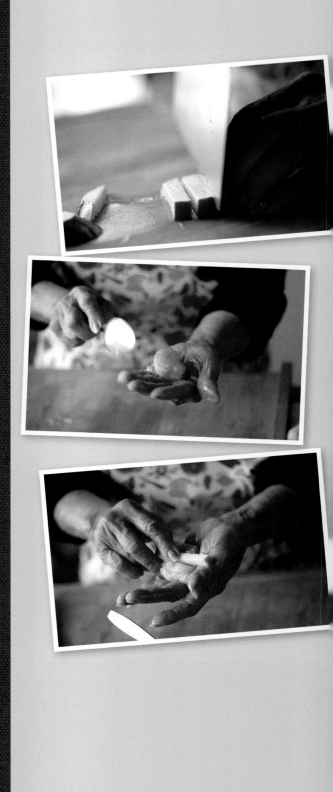

南乳炸青蠔

Deep Fried Mussel with Fermented Tarocurd

蔥燒三文魚頭
Fried Salmon Head with Chinese Leek

材料 Ingredients

三文魚頭　1 個
　京葱　1-2 條
　薑片　少許
　芫茜　1 棵
　乾葱　2 粒（切片）

1 salmon head
1-2 Chinese leeks
　sliced ginger
1 sprig coriander
2 cloves shallot (sliced)

調味料 Seasoning

　老抽　半湯匙
　生抽　半湯匙
　糖　半茶匙
胡椒粉　少許
　水　半杯

1/2 tbsp dark soy sauce
1/2 tbsp light soy sauce
　1/2 tsp sugar
　pepper
　1/2 cup water

做法 Method

1. 魚頭去鰓及鱗，清洗後用少許鹽、酒略醃，待用。
2. 京葱斜切長段，芫茜切段，待用。
3. 燒熱油將魚頭煎透（可走油），放入葱段、薑片、乾葱，灒酒，並加入調味料，燜煮至熟，放入芫茜即可上碟享用。

1. Scale salmon head and remove the gills. Rinse and marinate with salt and wine.
2. Cut Chinese leeks into long sections at an angle. Section coriander. Set aside.
3. Heat oil and fry salmon head thoroughly (or deep fry). Add Chinese leeks, ginger, shallot and sizzle with wine. Stir fry and add seasoning. Simmer until all ingredients cooked. Top with coriander. Serve.

京葱要炒透才會有香味，也可改用普通葱。三文魚頭有腥味，要煎透。

薄荷葉炒蝦球
Stir Fried Prawn with
Mint Leaves

材料 Ingredients

中蝦　1 斤
洋葱　4 両
紅椒　2 個
薄荷葉　3 棵
蒜茸　2 粒

600 g medium prawns
150 g onion
2 red bell peppers
3 sprigs mint
2 cloves garlic (grated)

醃料 Marinade

鹽　1/4 茶匙
胡椒粉、生粉　各少許

1/4 tsp salt
pepper
caltrop starch

調味料 Seasoning

魚露　1 1/2 湯匙
糖　半茶匙
麻油　少許
水　1 湯匙

1 1/2 tbsp fish sauce
1/2 tsp sugar
sesame oil
1 tbsp water

做法 Method

1. 中蝦剝殼取肉，挑腸，抹乾水分，放入醃料拌勻，泡油待用。
2. 洋葱切角，紅椒切塊，薄荷葉摘葉洗淨。
3. 用半湯匙油爆香蒜茸，將洋葱角、紅椒略炒，加入蝦球、調味料炒勻。
4. 將薄荷葉加入上項材料中輕手拌勻，即可上碟。

1. Shell prawns and devein. Wipe dry and mix well with marinade. Deep fry briefly with warm oil. Set aside.
2. Cut onion into wedges and cut red bell peppers into pieces. Take mint leaves only and rinse.
3. Fry garlic with 1/2 tbsp of oil until fragrant. Add onion and red bell peppers and stir fry. Add prawns and seasoning and stir fry well.
4. Add mint leaves and mix well. Serve.

薄荷葉後下，只取其香味。

苔條魚
Deep Fried Fish Fillet with Seaweed

寧波人最喜愛海苔，可炸後灑入花生中作小食。

材料 Ingredients

魚柳肉	約 8 両
麵粉	約 5 湯匙
雞蛋	1 個
水	適量
海苔末	1 湯匙

300 g fish fillet
5 tbsp flour
1 egg
water
1 tbsp dried seaweed flake

醃料 Marinade

鹽	半茶匙
酒	半湯匙
胡椒粉	少許

1/2 tsp salt
1/2 tbsp wine
pepper

做法 Method

1. 魚肉洗淨，吸乾水分，切成塊狀，放入醃料略醃片刻，待用。
2. 麵粉放大碗中，加入蛋和水，調勻成麵漿，加入海苔末拌勻。
3. 將魚塊沾上麵漿，放入熱油中炸至呈金黃色，撈起，瀝去油分上碟。可蘸點花椒鹽或甜酸汁享用。

1. Rinse fish and wipe dry. Cut into pieces, mix in marinade and set aside.
2. Put flour in a large bowl. Add whisked egg and water to make the batter. Mix in seaweed flake.
3. Coat fish fillet with the batter. Deep fry until golden brown. Drain and serve with pepper salt or sweet and sour sauce.

外省菜

我最初開始在烹飪中心教烹飪,是教上海菜,大約是 1979 年。那年代香港人還不是太熟悉上海菜,所以烹飪班很受歡迎,其實是錯打錯着,很是好笑。因為當時烹飪中心有八位導師,全是女性。我是新人,課程不能和其他同事有衝突,只能教上海菜,所幸我在上海長大,從小在父親身邊,去過很多省份,家中有廚師,父親亦講究飲食,所以一般的中菜也算有點認識,真想不到卻給我帶來了生路。所以,當我主持節目後,我也把一些特色的外省菜介紹給觀眾,不但引起他們的興趣,同時也使大家對外省菜有更深的瞭解,好像紅燒肋排,是標準的上海菜式。

當年在銅鑼灣近法國醫院有一間鄉村飯店,老闆是上海人,他開設一間很大規模的燈飾店,以賣水晶燈出名。他說,香港沒有一間做上海家鄉菜的店舖,就投資開了鄉村飯店,掌廚是他夫人,廚房全女班,紅燒肋排就是店中的名菜。那時,我還未入行,約是七零年的事。只因為大家都是上海人,老闆和我閒談,他並問我可有興趣加入他的廚房,我有少許心動,只因孩子們還小要照顧,婉拒這位老人家的好意。世事就是這般神奇,當年我與他都想不到,我會成為烹飪節目主持。您説,人生多有趣?!

苔條魚、紅燜鱔筒、白米蝦煮豆腐都是普通的上海家庭菜,但在店中卻很難嘗到了,真希望您們能學會。上海主婦習慣是中午與晚飯的菜式是一樣的,大多數是中午時煮好分成兩份,省時間、省柴火,所以上海女人大多數都有私己錢,午飯後一段時間可做自己的事。因此,在上海菜中多有紅燒燜煮的菜。現在看來很符合衛生條件,燜煮的菜需時較久,可把食材煮至熟透,會更健康好味。

芝麻蝦卷
Deep Fried Sesame Prawn

材料 Ingredients

中蝦　　12 隻
西芹絲、火腿絲　　各約 1 湯匙
白芝麻　　半杯
雞蛋　　1 個

12 medium prawns
1 tbsp shredded celery
1 tbsp shredded ham
1/2 cup white sesame
1 egg

醃料 Marinade

鹽　　1/4 茶匙
胡椒粉、粟粉　　各少許

1/4 tsp salt
pepper
cornstarch

火腿用中式火腿較有味，但會
較鹹，要切至極幼細絲。

做法 Method

1. 蝦剝殼切雙飛，洗淨拍扁，放入醃料拌勻。
2. 將每片蝦肉，捲入少許火腿絲、西芹絲，成球狀，沾上粟粉，再沾蛋汁少許，再沾上芝麻。
3. 將上項材料放入七成熱油中炸至呈金黃色，即可瀝去油分上碟。

1. Shell prawns and cut open. Rinse and pat. Mix well with marinade.
2. Roll ham and celery into each prawn, to make a ball shape. Coat in turn with cornstarch, whisked egg and sesame.
3. Deep fry prawn balls in hot oil until golden brown. Drain. Serve.

海鮮回鍋蛋
Fried Egg with Seafood

百花杞子釀翠環
Steamed Shrimp Stuffed Cucumber

素炒洋菜
Stir Fried Agar-Agar

118

材料 Ingredients

大菜	約半両
冬菇	6-8 朵
西芹	2 片
甘筍	半條

19 g agar-agar
6-8 dried shiitake mushrooms
2 pieces celery
1/2 carrot

調味料 Seasoning

生抽	約 1 湯匙
糖	半茶匙
麻油	少許

1 tbsp light soy sauce
1/2 tsp sugar
sesame oil

做法 Method

1. 大菜用清水浸軟，吸乾水分，剪成吋餘長段。
2. 冬菇浸透，去蒂洗淨，放入生抽、糖、油各少許，略蒸，切成粗絲。
3. 西芹、甘筍同切粗絲。
4. 燒熱油約 1 1/2 湯匙，放入西芹、甘筍等及少許鹽，潷入清水約 1/4 杯炒勻。
5. 加入冬菇、調味料，最後放入大菜炒拌均勻即成。

1. Soak agar-agar until soft. Wipe dry and cut into sections.
2. Soak dried shiitake mushrooms until soft. Remove stalk and rinse. Steam the mushrooms with light soy sauce, sugar and oil. Cut into thick shreds.
3. Cut celery and carrot into thick shreds.
4. Heat 1 1/2 tbsp of oil. Add celery, carrot and salt. Pour in 1/4 cup of water and stir well.
5. Add shiitake mushrooms and seasoning. Add agar-agar and stir fry well. Serve.

洋菜即是大菜，一般用作做果凍，但上海人喜炒後做涼菜，也可熱吃。洋菜要浸透、吸乾水、快炒，可口特別。

椒鹽肝醬脆豆腐
Deep Fried Stuffed Beancurd

材料 Ingredients

勝瓜　1條
甜梅菜　1棵（小）
碎肉　約2両
乾葱茸　1茶匙

1 luffa
1 small sprig sweet preserved mustard green
75 g ground pork
1 tsp grated shallot

調味料 Seasoning

生抽　半湯匙
生粉　半茶匙
水　1 1/2 湯匙

1/2 tbsp light soy sauce
1/2 tsp caltrop starch
1 1/2 tbsp water

做法 Method

1. 勝瓜刨皮留少許綠色，切滾刀塊，飛水後撈出，瀝乾水分上碟。
2. 梅菜洗淨，切小粒略剁碎，待用。
3. 調味料放入碎肉中拌勻。
4. 燒熱油1湯匙，爆炒碎肉、乾葱茸，加入梅菜炒勻。盛出放勝瓜面，隔水蒸至勝瓜熟即成。

1. Partly scrape off hard skin from luffa and cut into wedges. Blanch, drain and arrange on a plate.
2. Rinse preserved mustard green, cut into small dices and chop. Set aside.
3. Mix seasoning with ground pork.
4. Heat 1 tbsp of oil and stir fry ground pork and shallot until fragrant. Stir in preserved mustard green. Transfer on top of luffa. Steam until luffa cooked. Serve.

這道菜最重要是梅菜的味道，好的甜梅菜是不用加糖的，因此選擇梅菜時要先試味，否則會影響味道。

新老少平安
Steamed Mashed Beancurd with Dried Seafood

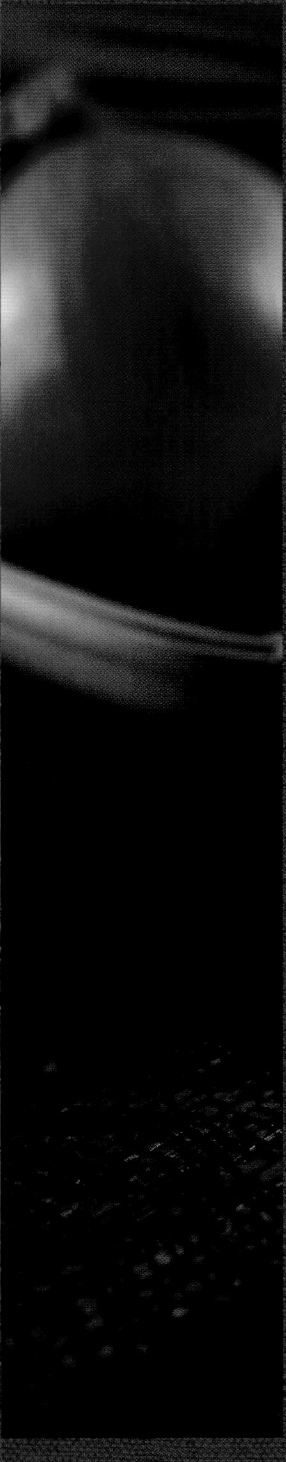

材 料 Ingredients

豆腐	1 件（大）
蝦米	半湯匙
瑤柱	1 粒
雞蛋、鹹蛋	各 1 個
葱粒	1 湯匙

1 large cube beancurd
1/2 tbsp dried shrimps
1 dried scallop
1 egg
1 salted egg
1 tbsp chopped spring onion

調 味 料 Seasoning

鹽	1/3 茶匙
生粉	1 1/2 茶匙
麻油	少許

1/3 tsp salt
1 1/2 tsp caltrop starch
sesame oil

做法 Method

1. 蝦米泡軟，瑤柱泡軟撕碎。
2. 豆腐放大碗中壓爛，放入雞蛋、鹹蛋白（鹹蛋黃切粒待用）。放入蝦米、瑤柱和調味料，一同攪拌均勻，放入深碟，鹹蛋黃放面，隔水蒸透，灑上葱粒，潷入滾油少許。可隨意淋上少許生抽，即可趁熱享用。

1. Soak dried shrimps until soft. Soak dried scallop until soft and tear into shreds.
2. Mash beancurd in a large bowl. Add egg and salted egg white. Chop salted egg yolk and set aside. Add dried shrimps, dried scallop and seasoning. Mix all ingredients together and transfer to a deep plate. Top with salted egg yolk. Steam thoroughly. Sprinkle with chopped spring onion and sizzle with hot oil. Add light soy sauce to suit your taste. Serve hot.

以前的老少平安只用少許鯪魚肉拌勻豆腐同蒸，此食譜改良後更為豐富，也可隨意加入少許魚肉。

材料 Ingredients

嫩西洋菜	約半斤
五香豆腐干	2 件
冬菇	8 朵
火腿茸	半湯匙

300 g young watercress
2 pieces dried flavoured beancurd
8 dried shiitake mushrooms
1/2 tbsp chopped ham

調味料 Seasoning

生抽	3/4 湯匙
糖	半茶匙
麻油	半湯匙

3/4 tbsp light soy sauce
1/2 tsp sugar
1/2 tbsp sesame oil

菜灼後要揸乾水分,方能入味,但又不能灼煮過久。

做法 Method

1. 西洋菜洗淨,放入滾水中灼熟,取出揸乾水分,切成極細小粒,放入少許麻油拌勻,待用。
2. 豆腐干飛水,切成極細小粒。
3. 冬菇浸透去蒂,放入少許生抽、糖,拌勻蒸熟,切成小粒。
4. 將西洋菜、豆腐干、冬菇同放大碗中,放入調味料拌勻,即可盛出上碟。將火腿茸放面,食時拌勻。

1. Rinse watercress. Boil until cooked, remove and squeeze excessive water. Chop finely. Mix in sesame oil. Set aside.
2. Blanch dried beancurd and chop finely.
3. Soak dried shiitake mushrooms until soft and remove stalks. Mix with light soy sauce, sugar and steam until cooked. Chop finely.
4. Combine watercress, dried beancurd and shiitake mushrooms in a large bowl. Mix in seasoning. Transfer to plate and top with chopped ham. Mix well before serving.

怎樣買菜、購物會慳功夫

做家庭主婦的朋友們說去市場買菜是難事，因為不知買甚麼好。的確，買家中伙食和雜用物品都需要一些技巧和學問，但又沒有學校可以攻讀，這就要靠思考、學習、經歷，再加上一些經驗才成。

去街市買菜、魚、肉類較新鮮、便宜、種類多。但亂買或一定要買某一種食材，或許都會花多了錢。我的方法是先在菜市場逛一下，觀察當天哪些材料多且便宜，然後構思早午兩餐的主菜，再配搭其他配菜或湯類，要有能下飯的菜，否則會不夠吃，如果是二人世界，要知道對方的口味，購買的量也要注意。現代人時間寶貴，選購可放入雪櫃保存的，就不需每天去街市了。此外有些肉類可燜煮熟保存一段時間。我一星期只買一次菜，蔬菜類可用報紙包裹放雪櫃下層，肉類放冰格，魚劏洗淨，灑少許鹽放碟內，蓋上保鮮紙放雪櫃上層，食用時再沖洗一次，抹乾即可烹調。我常將烏頭、黃花、鯰魚，甚至鯇魚腩、大魚等用此方法，保存二至三天都可以。菜類可先食用，瓜類、西蘭花、西芹、薯仔等都可保存。只要你肯試，走出第一步，就會令自己不太忙碌了，更可省錢，又吃得好。

去超市購物，因種類品種太多，常會見到就買，不知不覺裝滿購物車，超出預算。我的方法是將家中需購買的物品寫在紙上，然後照單執，不但不會超出預算且有帳可記，不要小看這些小錢，積少成多是名言。

作者簡介

　　方太（任利莎）原籍江蘇宜興，北京出生，十多歲到香港定居。

　　方太在 1979 年開始擔任電視節目的烹飪主持，深受觀眾歡迎；其後更出版食譜、飲食雜誌和擔任著名品牌的代言人，是香港人心目中的烹飪女王，她的烹飪魅力更風靡新加坡、馬來西亞的觀眾。

　　方太出版食譜迄今已逾 30 個年頭，菜式種類由南至北、中西菜式皆有，樂於新嘗試。除是次的《回味下飯菜》外，近年的著作有《粗茶淡飯・自煮同樂》、《團圓盛宴》、《福氣圓滿家常菜》和《你的看家菜》。

　　《粗茶淡飯。自煮同樂》是與三個女兒攜手合作出版，將方家四個女人，四種不同的飲食智慧，與讀者分享。在《團圓盛宴》中，方太與讀者分享做團圓菜的準備和心得，讓大家學懂如何有計劃地安排購物和烹調程序，從容入廚。方太認為有東西吃、有食物下鍋是福氣，要珍惜每一口，這是《福氣圓滿家常菜》的緣起，這書除有經典菜式、健康菜外，還有將剩菜新煮，創製成另一款美食。方太深信食譜能教讀者學會做菜，也能激發讀者的烹飪靈感，融會貫通另創新菜式，成為自己的看家菜，這亦是《你的看家菜》的由來，書內有醬料、小菜和點心，學會了就可大顯身手，贏盡喝采。

　　方太的食譜除美味、實際外，食譜內的引言、文章處處感受到她那濃濃的愛，以愛、情用心烹調的菜式，怎能不令你動容。

回味下飯菜 與您分享 最受歡迎的經典菜式
Those were the Tastes

作者	Author
方任利莎	Lisa Yam
策劃/編輯	Project Editor
	Catherine Tam
	Emily Luk
攝影	Photographer
	Imagine Union
美術統籌及設計	Art Direction & Design
	Amelia Loh
美術設計	Design
	Charlotte Chau

我嘉嘉唔係 咩都食㗎！

出版者　Publisher
Forms Kitchen
an imprint of Forms Publications (HK) Co. Ltd.
香港英皇道499號北角工業大廈18樓　18/F, North Point Industrial Building, 499 King's Road, Hong Kong
電話　Tel:　2138 7998
傳真　Fax:　2597 4003
網址　Web Site:　http://www.formspub.com
　　　　http://www.facebook.com/formspub
電郵　Email: marketing@formspub.com

發行者　Distributor
香港聯合書刊物流有限公司　SUP Publishing Logistics (HK) Ltd.
香港新界大埔汀麗路36號　3/F., C&C Building, 36 Ting Lai Road,
中華商務印刷大廈3字樓　Tai Po, N.T., Hong Kong
電話　Tel:　2150 2100
傳真　Fax:　2407 3062
電郵　Email: info@suplogistics.com.hk

承印者　Printer
中華商務彩色印刷有限公司　C & C Offset Printing Co., Ltd.

出版日期　Publishing Date
二〇一四年七月第一次印刷　First print in July 2014